OH BABY!

ALL ABOUT

BABY DOLPHINS

by Martha E. H. Rustad

PEBBLE
a capstone imprint

Pebble Emerge is published by Pebble, an imprint of Capstone.
1710 Roe Crest Drive
North Mankato, Minnesota 56003
www.capstonepub.com

Library of Congress Cataloging-in-Publication Data
Names: Rustad, Martha E. H. (Martha Elizabeth Hillman), 1975- author.
Title: All about baby dolphins / by Martha E.H. Rustad.
Description: North Mankato, Minnesota : Pebble, [2022] | Series: Oh baby! | Includes bibliographical references and index. | Audience: Ages 5-8 | Audience: Grades K-1 | Summary: "There's a new baby joining the pod. It's a dolphin calf! Learn all about baby dolphins, including what they eat, what they weigh, how they're raised, and how big they grow"—Provided by publisher.
Identifiers: LCCN 2021002863 (print) | LCCN 2021002864 (ebook) | ISBN 9781663907837 (hardcover) | ISBN 9781663907806 (pdf) | ISBN 9781663907820 (kindle edition)
Subjects: LCSH: Dolphins—Infancy—Juvenile literature.
Classification: LCC QL737.C432 R867 2022 (print) | LCC QL737.C432 (ebook) | DDC 599.5313/92—dc23
LC record available at https://lccn.loc.gov/2021002863
LC ebook record available at https://lccn.loc.gov/2021002864

Image Credits
Ardea Picture Library: Augusto Leandro Stanzani, 5; Getty Images: Rene Frederick, 15; Minden Pictures: Hiroya Minakuchi, 9, NPL/Pedro Narra, 6; Shutterstock: AlenKadr, 20 right, Andrea Izzotti, 19, cameilia, 20 left, Facanv, 12, Four Oaks, 13, Joost van Uffelen, 17, Monkey Business Images, 21, Rols Florent, 11, Tory Kallman, back cover, wildestanimal, 16; SuperStock: Masa Ushioda/age fotostock, cover

Editorial Credits
Editor: Alison Deering; Designer: Jennifer Bergstrom; Media Researcher: Tracy Cummins; Production Specialist: Tori Abrahahm

Printed in the United States 5680

Table of Contents

Words in **bold** are in the glossary.

A NEW DOLPHIN

Splash! Look at the baby dolphin. A baby dolphin is called a **calf**. It grew inside its mother for about one year.

The baby was born underwater. It must get to the surface to breathe. Dolphins breathe air through a **blowhole**.

dorsal fin

Dolphin babies look a lot like adults. But they are smaller. A newborn calf is about 3 feet (1 meter) long. Adults grow up to 12 feet (3.5 meters) long.

A newborn dolphin has a floppy **dorsal fin**. It was squished when the calf was inside its mother. The fin grows firmer after about two weeks.

A baby dolphin also has a belly button. This marks the spot where it got food while it was inside its mother.

The calf stays close to its mother. She feeds it milk from her body. The calf drinks milk for about two years. At about 6 months old, it also starts to eat fish.

belly button

LIFE UNDERWATER

Dolphins live in groups called **pods**. Each pod has five to 20 dolphins. They swim and hunt together.

Most pods are made up of females and their babies. The babies swim in the middle of the pod. This helps keep them safe.

Pods sometimes join together. They make larger groups called superpods. A superpod may have up to 1,000 dolphins.

Dolphins are born knowing how to swim. They move their tails up and down. This pushes their bodies through the water. They use fins and flippers to turn.

Adult dolphins are fast swimmers. They can swim up to 20 miles (32 kilometers) per hour.

DOLPHIN SOUNDS

Dolphins use sounds to **communicate**. They squeal, squeak, click, and whistle.

Each dolphin has its own whistle. This is like its name. A female dolphin teaches her whistle to her baby. She uses it to find her baby if they are apart.

Dolphins also use sound to find food. The sound bounces off fish. Dolphins can tell where the fish are.

Pods work together to hunt for food. They swim quickly around a school of fish. They push the fish into a ball. This makes it easier to grab the fish.

Dolphins bite the fish with their sharp teeth. They swallow their food whole.

GROWING UP

Dolphins stay with their mother until they are about 5 years old. They learn where to find fish. They learn how to hunt.

Dolphins also learn how to find their way in the big ocean. They learn how to stay safe.

A female dolphin has a baby when she is about 8 years old. Dolphins can live up to 50 years.

FIND YOUR FRIENDS

Dolphins use sound to find food under the water. Can you use sound to find your friends?

What You Need

- a container with a lid
- rocks or coins
- a blindfold
- an open place to play
- a group of friends

What You Do

1. Put the rocks or coins into the container. Shake it so everyone hears the sound.
2. One friend will be the dolphin. Put the blindfold on that friend. Everyone else is a fish.
3. Make a wide circle of fish around the dolphin. Quietly pass the container from fish to fish.
4. The dolphin listens carefully. If they hear a sound, they can point. That fish then becomes the dolphin.

Glossary

blowhole (BLOH-hohl)—a hole on the top of a dolphin's head; dolphins breathe air through a blowhole

calf (KAF)—a young dolphin

communicate (kuh-MYOO-nuh-kate)—to talk and pass along information

dorsal fin (DOR-suhl FIN)—a fin located on the back

pod (POD)—a group of dolphins; pods range from five to 20 dolphins

Read More

American Museum of Natural History. *Baby Dolphin's First Swim*. New York: Sterling Children's Books, 2017.

Leaf, Christina. *Dolphins*. Minneapolis: Bellwether Media, 2021.

Schuh, Mari. *The Supersmart Dolphin*. Minneapolis: Lerner Publications, 2019.

Internet Sites

Dolphin
animals.sandiegozoo.org/animals/dolphin

Meet the Dolphins
sheddaquarium.org/stories/meet-the-dolphins

National Geographic Kids: Bottlenose Dolphin
kids.nationalgeographic.com/animals/mammals/bottlenose-dolphin/

Index